THE ILLUSTRATED GUIDE TO COMMON FARM ANIMALS

A Vintage Field Companion to 50 Barnyard Favorites

by Luke Thomas
Published by The Terran Academy

Copyright & Notices

Published by The Terran Academy.

Disclaimer
This publication is intended for general educational and informational use. Descriptions, facts, and historical notes are presented for reference and enjoyment only. The material herein is not a substitute for professional expertise, certification, or safety instruction in any field. Readers are advised to exercise appropriate care, follow applicable laws and guidelines, and seek qualified guidance when necessary. Neither the author nor the publisher assumes responsibility for any outcome resulting from the use or interpretation of the information contained in this book.

Editing and design: Luke Thomas
Interior illustrations and art direction: The Terran Academy

First edition
Printed in the United States of America

10 9 8 7 6 5 4 3 2 1

ISBN (paperback): 978-1-970731-06-4

This book is dedicated to my children.
May you find joy in nature and peace in life.

A Note from the Author

Here at *The Terran Academy*, we are dedicated to the preservation of knowledge, history, art, and nature.

In this digital age, we are constantly overstimulated by bright lights and harsh colors as images and information flash endlessly across our screens. Our aim is to encourage readers to slow down and take time to explore the living world around them. We strive to promote a healthier environment for learning and to provide the perfect companion for doing just that.

Within our *Illustrated Guide* series, we have captured and presented the likeness of people, places, plants, animals, and human creations through beautiful vintage sketches—rendered in soft colors and intricate detail. Each book offers a wealth of information, including history, habitat, symbolism, and identification.

Our goal is to bring the nostalgia of simpler times into the modern era—those days when people paused to appreciate both creation and invention; when identifying trees and plants, along with their practical uses, was common knowledge; when recognizing the call of a particular bird or animal was second nature; and when our ability to live from the land was essential to our survival as a species.

These books are designed to invite discovery, to empower the curious beginner, and to aid in the teaching and appreciation of the natural world.

May these pages inspire you to look closer, wander slower, and see the beauty that blooms quietly all around us.

— Luke Thomas

TABLE OF CONTENTS

Horses & Ponies

Poultry & Fowl

Working & Companion Farm Animals

HOLSTEIN COW

Bos taurus

Identification:
The Holstein cow is large and stately, instantly recognized by its striking black-and-white or occasionally red-and-white patchwork coat. Mature females can weigh over 1,500 pounds, with a strong frame built for high milk production. Their clean, angular lines and deep udders distinguish them from beef breeds. Variations in markings are unique to each individual, almost like fingerprints among cattle.

Origin & History:
Holsteins trace their lineage to the northern provinces of the Netherlands, where selective breeding over two thousand years honed a dairy animal of remarkable yield. Dutch farmers prized them for both endurance and abundant milk, traits carried across the Atlantic in the mid-1800s when they were first imported to the United States. From those early herds, the Holstein became the foundation of modern dairy farming, admired for efficiency and gentle adaptability.

Habitat & Care:
Typically housed in open pastures with access to shade and shelter, Holsteins thrive in temperate climates but adjust well under careful management. Their comfort depends on steady feed, clean water, and routine milking schedules. Many farms provide soft bedding and climate-controlled barns to protect them from stress and heat.

Behavior & Temperament:
Holsteins are intelligent, calm, and social, often forming small herding groups. They respond well to routine and human attention, showing curiosity and mild-mannered patience during handling.

Agricultural Uses & Products:
Primarily bred for milk, Holsteins are unmatched in volume, producing the majority of the world's dairy supply. Their milk, rich in cream and protein, forms the base of countless cheeses, creams, and butters.

Cultural & Practical Notes:
A symbol of pastoral abundance, the Holstein's mottled hide has come to represent the dairy industry itself. Their dependable nature and enduring productivity make them one of humanity's most cherished agricultural companions.

JERSEY COW

Bos taurus

Identification:
The Jersey cow is small and fine-boned, with a soft fawn or light brown coat that may shade to gray or golden tan. Their large, dark eyes and gentle expression lend them an unmistakable charm. Many have a white muzzle or lighter underbelly, and their sleek hides show a delicate sheen. Though smaller than most dairy breeds, Jerseys are compact and efficient, bred for endurance rather than bulk.

Origin & History:
Originating from the Channel Island of Jersey, off the coast of France, this breed has been carefully maintained for centuries in relative isolation. By the early 1700s, Jerseys were already prized for their rich milk and even temperament. When introduced to England and later to North America in the 1800s, their reputation quickly spread. Farmers admired their thriftiness and the high butterfat content of their milk—traits that made them a cornerstone of traditional dairying.

Habitat & Care:
Jerseys thrive on modest pastures and adapt well to various climates, though they prefer mild weather. They are often kept in mixed systems—pastured by day and sheltered in open barns at night. Regular milking, balanced nutrition, and access to clean water ensure their continued productivity.

Behavior & Temperament:
Docile, intelligent, and friendly, Jersey cows form strong bonds with their handlers. Their alertness and curiosity make them quick to learn and easy to manage, a favorite among small farmers and family dairies.

Agricultural Uses & Products:
Renowned for their milk, which is rich in butterfat and protein, Jerseys are the preferred choice for producing cream, butter, and artisan cheeses.

Cultural & Practical Notes:
Beloved for both beauty and practicality, the Jersey remains a symbol of honest farming—gentle, efficient, and enduring, a true heritage breed of the dairy world.

ANGUS CATTLE

Bos taurus

Identification:
Angus cattle are medium-sized, muscular, and naturally polled (hornless), with a sleek, solid black or deep red coat. Their well-rounded bodies and fine bone structure make them both hardy and handsome. The black variety is most common, though the Red Angus, bred for heat tolerance, is equally admired in warmer climates. Their smooth hide and sturdy frame reflect centuries of adaptation to rugged pastures and varied weather.

Origin & History:
The Angus breed originated in northeastern Scotland, particularly in the counties of Aberdeen and Angus, where farmers developed them in the early 19th century from native Celtic stock. Known for their resilience and high-quality beef, they were exported widely during the 1800s, finding great success in North America. Selective breeding emphasized marbling, tenderness, and calm temperament, shaping the Angus into the premier beef breed known today.

Habitat & Care:
Angus cattle thrive in open pastures and temperate grasslands. They are adaptable to both cold northern regions and warmer southern ranges, provided with sufficient grazing and shelter during extremes. A diet rich in forage, clean water, and mineral supplements supports their robust growth and meat quality.

Behavior & Temperament:
Typically calm and intelligent, Angus cattle are social herd animals with a strong sense of order. They respond well to routine and are known for their gentle disposition, though bulls may display assertiveness when protecting their herd.

Agricultural Uses & Products:
Primarily raised for beef, Angus are prized for their well-marbled meat—rich, flavorful, and tender. Their hides are also valued for quality leather production.

Cultural & Practical Notes:
A symbol of strength and reliability, the Angus stands as a hallmark of modern ranching—a reminder of how good breeding and steady care can yield both beauty and bounty on the farm.

HERECORD CATTLE

HERREFORD CATTLE

Bos taurus

Identification:
Hereford cattle are medium to large in size, with a broad frame, muscular build, and a characteristic red body contrasted by a white face, crest, underbelly, and legs. Their thick hides and curly forehead hair help them withstand varying climates. Some are naturally horned, while others are polled through selective breeding. The rich chestnut-red color can vary slightly depending on region and lineage.

Origin & History:
The Hereford breed originated in Herefordshire, England, during the 18th century. Farmers sought a hardy, efficient animal capable of thriving on limited pasture while yielding quality beef. By the 1800s, Herefords had spread throughout Europe, the Americas, and Australia, where their resilience and docility made them a mainstay of open-range ranching. Over generations, selective breeding refined their beef quality and adaptability, cementing the breed as one of the world's most dependable cattle.

Habitat & Care:
Herefords are typically kept on open pasture and favor temperate to cool climates, though they adapt well to harsh or semi-arid environments. They require ample grazing land, clean water, and protection from heat stress. Their dense coats and strong constitution make them well suited to rugged terrain and seasonal weather shifts.

Behavior & Temperament:
Gentle, intelligent, and steady, Herefords are known for their mild manners and cooperative nature. They graze peacefully in herds and respond calmly to handling, making them a reliable choice for both small farms and large ranches.

Agricultural Uses & Products:
Primarily raised for beef, Herefords are prized for lean yet flavorful meat with excellent tenderness and marbling. Their hides are also used in leatherwork.

Cultural & Practical Notes:
Long regarded as a farmer's breed, the Hereford embodies thrift, endurance, and quiet strength—a living emblem of the pastoral traditions that shaped modern agriculture.

TEXAS LONGHORN

Bos taurus

Identification:
The Texas Longhorn is a striking breed, easily recognized by its sweeping horns that can extend over seven feet from tip to tip. It has a lean, muscular frame built for endurance and often displays a remarkable variety of coat colors—red, brindle, speckled, or patched in white and brown. Longhorns are medium to large in size, with sturdy legs and a strong neck, reflecting their rugged heritage. No two coats are exactly alike, lending each animal its own distinct character.

Origin & History:
Descended from cattle brought to the Americas by Spanish explorers in the late 1400s, the Longhorn adapted to the harsh conditions of the New World with extraordinary resilience. Over centuries of natural selection in the wilds of Mexico and the American Southwest, the breed developed its signature hardiness and distinctive horns. By the 1800s, the Texas Longhorn became a symbol of the American frontier and an essential part of the great cattle drives that helped shape the West.

Habitat & Care:
Longhorns thrive in open range environments and are exceptionally tolerant of heat and drought. They require minimal shelter and make efficient use of sparse forage, making them well suited to extensive grazing systems in semi-arid climates.

Behavior & Temperament:
Intelligent and self-reliant, Longhorns are alert and cautious but generally docile when handled with respect. Their keen instincts and strong herd loyalty reflect their centuries of survival on open land.

Agricultural Uses & Products:
Primarily raised for lean beef, Longhorns are valued for their flavor and natural disease resistance. Their hides and horns are also sought for crafts and decorative work.

Cultural & Practical Notes:
An enduring emblem of the American West, the Texas Longhorn stands for independence, endurance, and the untamed beauty of rural life—a living link to the pioneering spirit of the open range.

HIGHLAND CATTLE

Bos taurus

Identification:
Highland cattle are easily recognized by their long, shaggy coats and impressive curved horns. Their thick double-layered hair ranges in color from deep red and golden brown to black, dun, or silver, providing insulation against harsh weather. Compact yet sturdy, they have short legs and a broad, muscular frame designed for climbing rugged hillsides. Their gentle eyes peek through long forelocks that shield them from wind, rain, and biting insects.

Origin & History:
Native to the Scottish Highlands and western islands, this ancient breed dates back more than a thousand years. Bred by crofters to survive in cold, wet climates, Highland cattle became prized for their resilience, thrift, and high-quality beef. Over time, selective breeding emphasized their calm disposition and adaptability, allowing them to spread throughout Europe and beyond. By the 19th century, they were a staple of Scottish agriculture and a favorite export to North America and Australia.

Habitat & Care:
Highlands thrive in rough terrain and open pastures where other cattle might falter. Their thick coats eliminate the need for heavy shelter, even in winter. They prefer cool climates and benefit from rotational grazing to maintain healthy forage and hoof condition.

Behavior & Temperament:
Known for their calm, intelligent nature, Highland cattle are gentle yet hardy animals. They are social grazers with strong maternal instincts and rarely show aggression when handled with patience.

Agricultural Uses & Products:
Primarily raised for beef, their meat is lean, flavorful, and naturally lower in cholesterol. Their hides, with their long hair intact, are often used for rugs and decorative goods.

Cultural & Practical Notes:
A living emblem of the Scottish Highlands, these cattle embody endurance and grace in the face of adversity—a timeless reminder of the harmony between strength and gentleness in rural life.

BRAHMAN CATTLE

Bos indicus

Identification:
Brahman cattle are large, muscular animals easily identified by their distinctive hump over the shoulders, long drooping ears, and loose skin around the neck and dewlap. Their short, sleek coats vary from light gray to red or nearly black, often lightening around the face and underbelly. They possess strong legs, a high-set tail, and a noble bearing that reflects their tropical heritage.

Origin & History:
Descended from the sacred zebu cattle of India, Brahmans trace their ancestry back thousands of years. Revered in Hindu culture and bred for endurance in extreme heat, they were later imported to the Americas in the 19th century. American breeders combined several Indian strains to produce the modern Brahman, valued for its resistance to disease, parasites, and harsh climates. The breed soon became foundational to many hybrid beef lines throughout the southern United States and Latin America.

Habitat & Care:
Brahmans thrive in hot, humid environments where other breeds might struggle. Their loose skin aids heat dissipation and insect resistance. They require ample pasture, shade, and access to clean water, but need less intensive care than many European breeds thanks to their hardiness and efficient grazing habits.

Behavior & Temperament:
Alert and intelligent, Brahmans are often cautious but highly adaptable. With consistent, calm handling, they display a docile and cooperative temperament, particularly among cows and trained bulls.

Agricultural Uses & Products:
Primarily raised for beef, Brahman cattle yield lean, flavorful meat and serve as a genetic cornerstone in crossbreeding programs designed to improve heat tolerance and productivity.

Cultural & Practical Notes:
A symbol of strength and spiritual heritage, the Brahman represents the union of ancient endurance and modern adaptability—a living testament to mankind's partnership with nature across continents and centuries.

GUERNSEY COW

Bos taurus

Identification:
The Guernsey is a medium-sized dairy cow known for its warm fawn-and-white coat and gentle expression. It has a fine-boned yet sturdy frame, straight back, and well-developed udder suited for steady milk production. The breed's coloring can range from light golden tan to deep reddish-brown, often with white patches on the flanks or face. Guernseys are graceful in movement and carry an air of calm refinement that reflects their island heritage.

Origin & History:
Originating from the Isle of Guernsey in the English Channel, this breed was developed in the Middle Ages through selective breeding of French Norman and British stock. Isolated from the mainland, the Guernsey evolved into a distinct line renowned for producing rich, golden milk. By the late 1800s, Guernseys were exported to Britain and the Americas, where farmers prized them for their thriftiness and consistent yield.

Habitat & Care:
Guernseys thrive on pasture-based systems and moderate climates, tolerating both cool and warm conditions. They require balanced forage and clean shelter, with open fields allowing for natural grazing and exercise.

Behavior & Temperament:
These cows are known for their gentle, friendly disposition and are easily handled, making them favorites among family farms. They adapt well to human interaction and mixed herds alike.

Agricultural Uses & Products:
Primarily a dairy breed, the Guernsey produces milk with high butterfat and protein content, yielding a naturally golden hue due to beta-carotene. Their milk is prized for butter and artisan cheese production.

Cultural & Practical Notes:
Celebrated for both beauty and utility, the Guernsey remains a symbol of small-scale farming harmony—proof that patience, care, and good pasture can create nature's sweetest yield.

CHAROLAIS CATTLE

Bos taurus

Identification:
Charolais cattle are large, muscular animals noted for their creamy white to light beige coats and strong, broad frames. Their hides are short-haired and sleek, sometimes showing a soft silvery sheen in sunlight. Both bulls and cows may be naturally horned or polled, and they stand tall with deep chests and powerful hindquarters, giving them a striking and stately appearance.

Origin & History:
The Charolais breed originated in the Charolles and Nievre regions of France, where it was first developed during the 16th and 17th centuries. Initially used as draft animals, they were later bred for beef production, valued for their rapid growth and abundant muscle. Charolais cattle were introduced to North America in the 1930s, where they quickly gained popularity for improving size and meat yield in crossbreeding programs.

Habitat & Care:
Hardy and adaptable, Charolais thrive on open pastures and moderate to cool climates. They benefit from ample grazing space and dry bedding during wetter months. Though they tolerate heat, shaded areas and fresh water are essential for comfort and optimal performance.

Behavior & Temperament:
Generally calm and manageable, Charolais cattle are intelligent and curious, though bulls can be protective of their herd. When handled regularly and respectfully, they respond well to quiet management and structured routines.

Agricultural Uses & Products:
Primarily raised for beef, Charolais cattle are renowned for producing lean, tender meat with excellent carcass yield. They are also a cornerstone in hybrid breeding, improving muscle and efficiency in mixed herds worldwide.

Cultural & Practical Notes:
From the fertile fields of Burgundy to modern ranches across the globe, the Charolais remains a symbol of strength, refinement, and agricultural progress—a quiet reminder of the enduring art of good breeding and good stewardship.

SIMMENTAL CATTLE

Bos taurus

Identification:
Simmental cattle are among the oldest and most visually striking of European breeds, recognized for their large frames, deep chests, and strong, balanced builds. Their coats range from pale gold to deep red with distinctive white faces, legs, and underbellies. The breed is often horned, though many modern herds are polled, and individuals vary slightly in color and markings depending on regional bloodlines.

Origin & History:
The Simmental originated in the Simme Valley of Switzerland, where it was prized as a dual-purpose breed for both milk and meat production. Historical records trace its lineage to the Middle Ages, when farmers selected for strength, fertility, and gentle disposition. Over time, Simmental cattle spread across Europe and into the Americas, adapting to varied climates while maintaining their signature productivity and adaptability.

Habitat & Care:
These cattle thrive in cool to moderate climates and are typically raised on open pastureland. They do best with ample grazing space, clean water, and sturdy shelter during harsher seasons. Their thick hides provide natural resistance to colder conditions, and they are efficient converters of forage into weight and milk.

Behavior & Temperament:
Simmentals are known for their intelligence, calm nature, and strong maternal instincts. They bond closely within herds and are easily managed when treated with patience and consistency.

Agricultural Uses & Products:
Highly versatile, Simmentals provide both rich milk and superior beef. Their dual-purpose lineage makes them invaluable to mixed-use farms, yielding high-quality dairy and lean, flavorful meat.

Cultural & Practical Notes:
Long associated with reliability and abundance, the Simmental remains a cornerstone of traditional agriculture—a reminder that balance between strength and gentleness is the farmer's finest ally.

MERINO SHEEP

Ovis aries

Identification:
The Merino is a medium-sized sheep with a sturdy yet refined frame, covered in exceptionally dense, fine wool that forms soft folds around the neck and shoulders. Its fleece ranges from creamy white to pale ivory, prized for its silky texture and elasticity. Rams often bear large, spiraled horns, while ewes are typically polled. Across regions, the thickness of the fleece and the degree of skin folding may vary depending on climate and breeding line.

Origin & History:
Originating in Spain around the 12th century, the Merino became the foundation of Europe's wool trade and a symbol of national wealth. For centuries, Spain tightly guarded the breed, exporting it only under royal decree. By the 18th century, Merinos reached Australia and New Zealand, where careful selection and adaptation produced some of the world's finest wool. Today, their lineage underpins global textile industries and the art of sustainable fiber farming.

Habitat & Care:
Merinos thrive in temperate, dry pastures and open rangelands. They require regular shearing—typically once or twice a year—to maintain comfort and prevent overheating. Clean shelter, ample grazing, and parasite management are essential for flock health.

Behavior & Temperament:
These sheep are gentle, intelligent, and highly social, often moving as a coordinated flock. They respond well to calm handling and regular routine.

Agricultural Uses & Products:
The Merino is primarily raised for its fine-grade wool, which is used in high-quality clothing and textiles. Its meat, though secondary, is tender and flavorful.

Cultural & Practical Notes:
Renowned as the "gold standard" of wool, the Merino embodies patience, craftsmanship, and the enduring partnership between shepherd and flock—a living testament to centuries of careful stewardship.

SUFFOLK SHEEP

Ovis aries

Identification:
The Suffolk is a robust, broad-framed sheep, admired for its muscular build and graceful carriage. Its most recognizable feature is the striking contrast between its creamy white wool and jet-black face and legs, which are smooth and free of fleece. The head carries an alert, noble expression with long ears and a broad muzzle. Suffolks are naturally polled and stand tall and deep-bodied, with variations in size and wool density seen among flocks raised in different climates or pasture systems.

Origin & History:
This celebrated breed traces its roots to southeastern England during the late 18th century, born from the crossing of old Norfolk Horn ewes with Southdown rams. The Suffolk soon gained renown for its vigor and meat quality, and by the Victorian era, it became a favored sheep throughout Britain. As agriculture expanded across continents, the breed followed, establishing strong footholds in North America, Australia, and beyond. Careful selection refined its hardy frame and dependable growth, shaping the Suffolk into one of the most successful meat sheep in modern husbandry.

Habitat & Care:
Suffolks flourish on open pastures with firm footing and ample forage. They tolerate varied climates well but benefit from dry shelter, clean bedding, and regular hoof and parasite care. Rotational grazing and balanced nutrition support healthy development and fleece condition.

Behavior & Temperament:
Known for their calm disposition and keen awareness, Suffolks are social flock animals that bond readily with companions and adapt well to human handling.

Agricultural Uses & Products:
Primarily raised for premium lamb and mutton, the Suffolk also produces a modest fleece of medium wool suitable for durable textiles.

Cultural & Practical Notes:
From English heaths to American family farms, the Suffolk stands as a symbol of dependable strength and pastoral tradition—an enduring partner in the quiet rhythm of rural life.

DORSET SHEEP

Ovis aries

Identification:
Dorset sheep are medium to large in size, with a strong, blocky frame and dense, creamy-white fleece. Their faces and legs are also white and free of wool, with a pink nose and bright, inquisitive eyes. Both horned and polled varieties exist, the former carrying symmetrical, curling horns that lend a distinguished appearance. The breed's wool is of medium grade—soft, springy, and well-suited to hand spinning and knitwear. Minor variations in fleece weight and body size occur depending on regional breeding and climate.

Origin & History:
The Dorset originated in southern England's Dorset Downs, where early shepherds bred hardy local ewes with Spanish Merino and Wiltshire rams. By the 17th century, the breed was prized for its dual-purpose nature—producing both excellent meat and abundant wool. The Dorset's unique ability to breed out of season distinguished it from most sheep, leading to widespread adoption in the United States and other temperate regions. Selective breeding in the 20th century refined its conformation and mothering traits, establishing it as one of the most versatile sheep in agriculture.

Habitat & Care:
Dorsets thrive on rolling pastures and tolerate a variety of climates, though they prefer well-drained ground and moderate shelter in harsh weather. They require shearing once or twice yearly, routine hoof care, and balanced nutrition to sustain steady growth and lambing cycles.

Behavior & Temperament:
Gentle and intelligent, Dorsets are highly social and easily handled. They are attentive mothers, often producing twins or triplets, and adapt well to both small farms and larger operations.

Agricultural Uses & Products:
Dorsets are valued for their tender, flavorful lamb and quality fleece. Their year-round breeding capability allows for consistent production, making them ideal for sustainable farming.

Cultural & Practical Notes:
Beloved for their calm nature and dependable yield, the Dorset embodies the pastoral ideal—a steadfast companion in the timeless rhythm of field and flock.

JACOB SHEEP

Ovis aries

Identification:
The Jacob Sheep is a striking and ancient breed, easily recognized by its piebald coat of white fleece boldly splashed with black or dark brown. Most individuals carry two, four, or even six symmetrical horns, lending them a regal and almost biblical appearance. Their medium build, lean frame, and fine-boned legs give them a graceful stance uncommon among heavier wool breeds. The fleece is soft and medium-length, yielding beautiful natural-spotted wool for handspinners. Variations in horn number and pattern make each Jacob distinct, a living example of nature's artistry.

Origin & History:
Named after the biblical story of Jacob's speckled flock, the breed is believed to have ancient Middle Eastern origins, later spreading across North Africa and Europe. By the 17th century, Jacob sheep were kept on English estates as ornamental and productive flock animals. Their hardy nature and distinctive looks preserved them through centuries of agricultural change. Modern conservation efforts have safeguarded the breed's purity, celebrating it as one of the oldest surviving examples of primitive sheep stock.

Habitat & Care:
Jacobs thrive in temperate pastures and rugged terrain. They are highly self-sufficient and resist many common ailments, needing little more than rotational grazing, dry shelter, and yearly shearing. Their light frame allows efficient grazing on sparse forage, ideal for small or low-input farms.

Behavior & Temperament:
Alert, curious, and independent, Jacobs are intelligent flock animals with gentle manners when handled calmly. They bond readily with keepers and adapt well to diverse farm settings.

Agricultural Uses & Products:
Valued for both meat and fleece, Jacob wool—naturally colored and soft—is prized by artisans. Their lean, flavorful meat is another hallmark of the breed's utility.

Cultural & Practical Notes:
The Jacob Sheep stands as a symbol of heritage husbandry—ornamental yet practical, resilient yet gentle—bridging the ancient pastoral past with the sustainable farms of today.

BOER GOAT

Capra hircus

Identification:
The Boer Goat is a large, muscular breed distinguished by its white body and deep chestnut-red head and neck. Its pendulous ears and gentle, convex facial profile give it a calm, noble appearance. Both sexes are typically horned, with sturdy, curved horns and a strong, arched back. The coat is short and sleek, suited to warm climates, though occasional variations appear in pattern and shade depending on regional breeding.

Origin & History:
The Boer Goat originated in South Africa in the early 20th century, developed by Dutch settlers who crossbred indigenous goats with imported European and Indian stock. The goal was a hardy animal with rapid growth, excellent fertility, and high-quality meat. "Boer," meaning "farmer" in Dutch, reflects its agricultural roots. Over time, the breed spread across continents, becoming the world's leading meat goat through its adaptability and productivity. Selective breeding refined its conformation and feed efficiency, securing its role in modern livestock farming.

Habitat & Care:
Boer goats thrive in dry, open pastures and are remarkably resilient to heat and drought. They require sturdy fencing, simple shelter from wind and rain, and regular hoof trimming. Their browsing habits help manage weeds and shrubs, making them valuable in land stewardship.

Behavior & Temperament:
Docile and intelligent, Boers are social herd animals that bond closely with others and respond well to gentle handling. Nannies are attentive mothers with strong maternal instincts.

Agricultural Uses & Products:
Primarily raised for their tender, lean meat, Boer goats also contribute hides and limited milk. Their rapid growth and efficient feed conversion make them central to sustainable small-scale farming.

Cultural & Practical Notes:
A symbol of strength and abundance, the Boer Goat embodies the spirit of practical husbandry—an enduring partner in the balance between pasture, people, and plenty.

NUBIAN GOAT

Capra hircus

Identification:
The Nubian Goat is a tall, elegantly built breed known for its distinctive Roman nose and long, pendulous ears that hang gracefully below the jawline. Its coat can range from solid white to jet black, or display a wide array of mottled and speckled patterns in shades of tan, chestnut, or gray. Both sexes may bear short, curved horns. With their proud carriage and expressive faces, Nubians are as ornamental as they are practical, their appearance varying subtly with climate and lineage.

Origin & History:
The Nubian traces its roots to the arid regions of North Africa and the Middle East, where desert-adapted goats were prized for their stamina and milk yield. British breeders in the 19th century crossed these hardy African goats with Indian and English dairy lines, creating the Anglo-Nubian—a refined breed noted for rich milk and graceful form. Exported worldwide, the Nubian became one of the most recognizable dairy goats, celebrated for both productivity and beauty.

Habitat & Care:
Adapted to hot, dry climates, Nubians thrive in open pastures with shade and clean water readily available. They prefer roomy shelters and good ventilation, requiring regular grooming, hoof care, and balanced nutrition to maintain high milk production.

Behavior & Temperament:
Affectionate and intelligent, Nubians are known for their vocal personalities and strong social bonds. They are curious, alert, and often follow their keepers with doglike loyalty.

Agricultural Uses & Products:
Nubians are prized for their rich, high-butterfat milk, ideal for cheese and soap making. Their gentle nature also makes them favored companion and show animals.

Cultural & Practical Notes:
Graceful and enduring, the Nubian Goat stands as a symbol of refinement and resilience—bridging centuries of pastoral tradition with the modern art of sustainable living.

ALPINE GOAT

Capra hircus

Identification:
The Alpine Goat is a medium-to-large dairy breed, athletic in build and notable for its upright ears, straight profile, and elegant stance. Its coat varies widely in color—ranging from pure white to shades of fawn, gray, brown, or black—often marked with distinctive facial stripes or contrasting legs and belly. Both sexes may carry short, backward-curving horns. Adaptable and sure-footed, the Alpine's clean lines and alert expression reflect its mountain heritage.

Origin & History:
The breed originated in the rugged mountain pastures of the French and Swiss Alps, where it was developed by generations of herdsmen selecting for stamina, hardiness, and abundant milk yield. Introduced to North America in the early 20th century, Alpine goats quickly gained favor for their productivity and adaptability to diverse climates. Their ancestry ties them to centuries of European dairy tradition, where they were valued as essential companions to shepherds traversing steep alpine meadows.

Habitat & Care:
Alpines thrive in hilly or open pasture environments with good forage variety. They tolerate both cool and temperate conditions, though they appreciate dry bedding and well-ventilated shelters. Regular grooming and hoof care are essential, as is access to clean water and mineral supplements to support their high energy and milk output.

Behavior & Temperament:
Curious, intelligent, and agile, Alpines are confident climbers with a spirited yet manageable temperament. They are social animals that thrive in groups and interact readily with humans.

Agricultural Uses & Products:
Primarily valued for their milk, Alpines produce large quantities with a mild, sweet flavor ideal for fresh dairy and cheese making.

Cultural & Practical Notes:
A symbol of mountain endurance and quiet industry, the Alpine Goat remains a cornerstone of traditional dairying—nimble, efficient, and ever faithful to the small farm's rhythm.

ANGORA GOAT

Capra hircus

Identification:
The Angora Goat is a medium-sized, delicately built animal distinguished by its long, lustrous ringlets of fine, silky fleece known as mohair. Its coat ranges from gleaming white to shades of silver-gray, with occasional colored strains in modern herds. Both sexes bear gently curved horns, and their narrow faces, pendulous ears, and calm eyes lend an air of refinement. Shorn twice yearly, Angoras appear almost transformed between coats—a testament to the abundance of their fleece.

Origin & History:
The Angora traces its lineage to the Ankara region of ancient Turkey, where its prized mohair was traded along Silk Road routes as early as the 13th century. Revered by sultans and merchants alike, the breed eventually spread to Europe and the Americas, finding a stronghold in the dry plains of Texas and South Africa. Selective breeding refined the uniformity and sheen of its fleece, making it a staple of the luxury textile trade for centuries.

Habitat & Care:
Angoras prefer dry, temperate regions with open pastures. Their delicate fleece requires protection from damp conditions, so clean, well-drained shelters are essential. Regular shearing, grooming, and mineral-rich diets maintain coat health and prevent matting or loss of luster.

Behavior & Temperament:
Gentle and unhurried, Angoras are docile by nature but less hardy than common meat goats. They thrive in calm, attentive environments and bond easily within their flock.

Agricultural Uses & Products:
Renowned for mohair production, Angoras yield several pounds of fiber per shearing—prized for its gloss, durability, and natural dye affinity.

Cultural & Practical Notes:
Once the treasure of ancient looms, the Angora Goat endures as a living thread between pastoral craft and timeless elegance—a quiet marvel of patience and beauty in motion.

NIGERIAN DWARF GOAT

Capra hircus

Identification:

The Nigerian Dwarf Goat is a miniature dairy breed of delicate proportions, standing barely two feet tall at the shoulder yet carrying a proud and graceful bearing. Its coat is short, sleek, and found in nearly every color and pattern—black, gold, chocolate, white, and spotted varieties abound. Both sexes may bear small, curved horns, and their bright eyes and upright ears give them an alert, lively expression.

Origin & History:

Originating from the West African dwarf goat, this breed was first brought to the Americas aboard trading ships in the early 20th century. Initially kept in zoos and private collections, they were later selectively bred for gentler temperaments and superior dairy traits. Their milk, remarkably rich and sweet, earned them favor among small farmers and homesteaders, transforming a once-exotic novelty into one of the most beloved miniature livestock breeds.

Habitat & Care:

Compact and hardy, Nigerian Dwarfs adapt well to both rural pastures and small farmyards. They thrive in moderate climates with shelter from rain and cold winds, requiring secure fencing, clean bedding, and a diet balanced between browse, hay, and grains.

Behavior & Temperament:

Intelligent, affectionate, and full of personality, these goats are social and playful. They bond easily with humans and herd mates, often responding to names and seeking attention.

Agricultural Uses & Products:

Though small, Nigerian Dwarfs produce rich, high-butterfat milk ideal for cheese and soaps. Their gentle natures also make them favorites for youth farming and therapy programs.

Cultural & Practical Notes:

A true "little giant" of the barnyard, the Nigerian Dwarf Goat embodies the charm of miniature livestock—offering abundance, companionship, and joy well beyond its size.

CASHMERE GOAT

Capra hircus

Identification:
The Cashmere Goat is a medium-sized animal with a sturdy frame and a dual coat consisting of a coarse outer layer and an exceptionally fine, downy undercoat—the source of luxurious cashmere fiber. Colors range from white and cream to shades of gray, brown, or black, depending on regional strains. Both sexes often have sweeping, curved horns, and their calm, intelligent eyes are framed by a thick mane that hints at their cold-climate origins.

Origin & History:
Native to the high plateaus of Central Asia—particularly Mongolia, Tibet, and Kashmir—the Cashmere Goat has been bred for thousands of years for its soft, insulating fiber. Ancient traders prized its wool for its unmatched warmth and lightness, transporting it along the Silk Road to markets in Persia, India, and Europe. Modern Cashmere goats are derived from various regional lines, carefully bred for fiber yield and fineness while preserving their rugged, mountain-born resilience.

Habitat & Care:
These goats thrive in cool, arid regions with open grazing lands. Their thick coats protect them from harsh winters, but they need shade and clean water in summer. Annual combing or shearing during molting season collects their valuable undercoat without harm.

Behavior & Temperament:
Cashmere goats are hardy yet gentle, known for independence balanced by quiet sociability. They are intelligent grazers, well-suited to diverse pastures.

Agricultural Uses & Products:
Renowned for producing cashmere fiber—one of the finest natural materials in the world—these goats also provide meat and milk in some regions.

Cultural & Practical Notes:
Treasured by shepherds and artisans alike, the Cashmere Goat remains a living emblem of patient craft—turning mountain wind and time into warmth, softness, and grace.

YORKSHIRE PIG

Sus scrofa domesticus

Identification:
The Yorkshire Pig is a large, strong-bodied breed distinguished by its clean white coat, upright ears, and long, deep torso. Its skin is smooth and nearly hairless, with a broad back and well-developed hams—a physique built for both stamina and productivity. The breed's even proportions and bright, alert expression reflect its reputation for hardiness and efficiency, making it one of the most recognizable swine on modern farms.

Origin & History:
Originating in Yorkshire, England, during the early 19th century, the breed was developed through selective crossing of local white pigs prized for size and mothering ability. Its adaptability and consistent performance soon led to export across Europe and North America. In the United States, where it is often called the "Large White," the Yorkshire quickly became a cornerstone of commercial pork production. Its refined frame and dependable yield continue to influence many hybrid lines used today.

Habitat & Care:
Yorkshires thrive in managed outdoor pastures or well-ventilated barns with dry bedding. They are tolerant of varied climates but benefit from shade and shelter to protect their sensitive light-colored skin. Clean water and balanced feed are essential to sustain their rapid growth and reproductive vigor.

Behavior & Temperament:
These pigs are intelligent, curious, and highly social. They interact readily with humans and herd mates, showing a calm but active disposition.

Agricultural Uses & Products:
Renowned for lean, high-quality pork, Yorkshires are also valued for prolific breeding and strong maternal instincts.

Cultural & Practical Notes:
From English farms to American heartlands, the Yorkshire Pig stands as a quiet emblem of agricultural progress—faithful, efficient, and ever at home in the rhythm of the barnyard.

BERKSHIRE PIG

Sus scrofa domesticus

Identification:
The Berkshire Pig is a medium to large breed known for its glossy black coat accented by distinctive white points on the legs, snout, and tail. It has a compact, muscular body, short legs, and an upturned snout that gives it a cheerful, alert appearance. Its smooth skin and dense, fine hair reflect careful breeding for both beauty and utility. Slight variations in shade and body mass appear among regional lines, but all share the breed's signature sturdiness and balance.

Origin & History:
The Berkshire breed traces its roots to the shire of Berkshire in England, where it was first developed in the 1600s. Early farmers crossed native pigs with imported Asian stock to create a hardier, faster-growing animal prized for its marbled, flavorful meat. The breed's excellence in pork quality earned royal favor—legend has it that even the House of Windsor maintained a herd. Introduced to America in the early 19th century, Berkshires soon became a foundation breed for premium pork production.

Habitat & Care:
Berkshires thrive in open pastures or clean pens with access to shade and fresh bedding. They are well adapted to temperate climates and respond favorably to balanced diets rich in grains and greens.

Behavior & Temperament:
Calm, intelligent, and sociable, Berkshires are known for their friendly nature and ease of handling. They display curiosity and strong social bonds within their groups.

Agricultural Uses & Products:
Primarily raised for their richly marbled pork, Berkshires produce tender, flavorful meat prized by chefs and small farmers alike.

Cultural & Practical Notes:
A living link to England's rural past, the Berkshire Pig remains a symbol of tradition, quality, and the enduring artistry of heritage farming.

HAMPSHIRE PIG

Sus scrofa domesticus

Identification:
The Hampshire Pig is easily recognized by its striking black body crossed with a broad white belt that encircles the shoulders and front legs. Medium to large in size, it has a lean, muscular frame, upright ears, and a long, narrow face. The coat is sleek and fine, reflecting the breed's vitality and health. Hampshires vary slightly in body length and belt width among regional strains but consistently display an athletic, well-proportioned build.

Origin & History:
Hampshires are among the oldest recorded pig breeds, believed to have originated in southern England centuries ago. Early stock was exported to America around 1830, where careful selection refined the breed for its efficiency and quality meat. Its English ancestors were valued for hardiness and flavor, while American breeders emphasized lean growth and strong mothering traits, producing the balanced and dependable Hampshire known today.

Habitat & Care:
This breed thrives in moderate climates and adapts well to both pasture and confined housing. Clean, dry pens and access to forage or quality grain help maintain muscle tone and steady growth. Hampshires tolerate a range of weather but benefit from shade and wind protection.

Behavior & Temperament:
Hampshire pigs are active, intelligent, and even-tempered. They are strong foragers with social instincts that favor herd living and gentle interaction with caretakers.

Agricultural Uses & Products:
Prized for their lean, flavorful pork, Hampshires are widely used in breeding programs to enhance carcass quality and growth efficiency in other lines.

Cultural & Practical Notes:
A hallmark of reliability and refinement, the Hampshire Pig stands as a testament to the farmer's craft—an enduring symbol of balance between beauty, strength, and purpose.

DUROC PIG

Sus scrofa domesticus

Identification:
The Duroc Pig is a robust, medium-to-large breed known for its rich reddish-brown coat, which ranges from deep mahogany to light gold. It has a well-muscled frame, slightly drooping ears, and a strong, arched back that hints at both power and endurance. The breed's smooth skin and uniform color give it a striking appearance in the barnyard. Minor variations in shade and size appear depending on local breeding lines, but all Durocs share their signature vigor and graceful build.

Origin & History:
The Duroc traces its roots to early American pig stock, likely a blend of red pigs from the northeastern United States and imported European breeds. By the mid-1800s, farmers in New York and New Jersey refined these animals into a consistent line renowned for meat quality and adaptability. The breed took its name from a favored stallion of the era, symbolizing its growing reputation in American agriculture. Durocs became a mainstay in the Midwest, where their hardy constitution suited both open pasture and confinement systems.

Habitat & Care:
Durocs are resilient and thrive in diverse climates. They do best in clean, shaded pens or open pastures with access to water and balanced feed. Their tough skin and even temperament make them easy keepers for both small farms and large operations.

Behavior & Temperament:
Calm yet active, Durocs are intelligent, inquisitive, and notably docile. They form social bonds within herds and show strong maternal instincts.

Agricultural Uses & Products:
Primarily raised for high-quality pork, Durocs are prized for marbling and tenderness, contributing genetics to many modern hybrid lines.

Cultural & Practical Notes:
A symbol of American farm ingenuity, the Duroc stands as a dependable producer and a reminder of the quiet artistry behind generations of careful breeding.

TAMWORTH PIG

Sus scrofa domesticus

Identification:
The Tamworth Pig is a medium-sized, long-bodied breed noted for its rich, reddish-gold coat and fine bristles that gleam in sunlight. It has a narrow head, erect ears, and a strong, slightly arched back suited for foraging. Its lean frame and long snout distinguish it from heavier commercial breeds. Though shades of red vary from deep copper to light ginger, all Tamworths share the same rustic elegance and agility that mark their heritage as true outdoor pigs.

Origin & History:
One of Britain's oldest and purest pig breeds, the Tamworth originated in Staffordshire, England, during the early 1800s. Developed from hardy forest pigs, it was refined by crossing native stock with Irish breeds, resulting in an animal well suited to roaming wooded pastures. The Tamworth's ability to thrive on rough forage made it invaluable to small farms. Imported to North America in the late 19th century, it became admired for its distinctive bacon and old-fashioned hardiness.

Habitat & Care:
Tamworths flourish in open, grassy environments where they can root and explore. They require shelter from harsh weather, clean bedding, and ample room to roam. Their thick skin and coat provide resilience to cold climates, though shade and wallows are essential in warmer regions.

Behavior & Temperament:
Active and intelligent, Tamworths are curious foragers with gentle dispositions. They form social herds and exhibit strong maternal care.

Agricultural Uses & Products:
Tamworths are renowned for their lean, flavorful bacon and high-quality pork. Their grazing efficiency and minimal input needs make them ideal for pasture-based and heritage farms.

Cultural & Practical Notes:
A living relic of England's woodland past, the Tamworth endures as a symbol of self-sufficiency — a breed that connects modern farms to their agrarian roots.

CLYDESDALE

Equus ferus caballus

Identification:
The Clydesdale is a grand and powerful draft horse, easily recognized by its tall stature, broad chest, and long, feathered legs. Typically standing between 16 and 18 hands high, it displays a proud bearing with a high-set neck and expressive eyes. The coat is most often bay with white markings on the face and legs, though black, brown, and roan are also seen. Its white feathering and rhythmic gait give the breed a noble, almost ceremonial presence.

Origin & History:
Originating in the Clyde River Valley of Scotland during the 18th century, the Clydesdale was bred from native Scottish horses crossed with large Flemish draft stock. Its early purpose was to haul coal and farm produce, yet its beauty and strength soon earned it fame beyond the fields. Exported throughout the British Empire, the breed became synonymous with reliability and grandeur, later immortalized by brewery teams and parades.

Habitat & Care:
Clydesdales require spacious pastures and sturdy stables to accommodate their size. They thrive in temperate climates with regular grooming to maintain their flowing leg feathering. Adequate feed and gentle exercise preserve their stamina and impressive musculature.

Behavior & Temperament:
Gentle giants by nature, Clydesdales are patient, intelligent, and eager to work. Their calm temperament and steady step make them well suited for harness, riding, or ceremonial use.

Agricultural Uses & Products:
Traditionally valued for draft labor, they remain icons of strength in logging, farming, and exhibition.

Cultural & Practical Notes:
The Clydesdale embodies both power and grace — a living monument to the era when horsepower meant life itself, and the bond between farmer and field was forged in mutual effort.

AMERICAN QUARTER HORSE

Equus ferus caballus

Identification:
The American Quarter Horse is a compact, muscular breed known for its powerful hindquarters and calm, alert expression. Standing between 14 and 16 hands high, it exhibits a balanced frame ideal for bursts of speed and agility. Coat colors vary widely — from sorrel and chestnut to palomino, bay, and black — often accented by subtle facial or leg markings. Its sturdy build and kind eyes reflect both strength and good nature, traits long admired on American farms and ranches.

Origin & History:
This breed traces its lineage to colonial America, where settlers crossed English thoroughbreds with hardy native horses. Named for its unmatched sprint over a quarter-mile, the Quarter Horse became essential to frontier life, cattle work, and early racing. Its intelligence and adaptability made it the foundation of the American ranching tradition. By the 20th century, the breed had achieved global recognition as a symbol of the Western spirit.

Habitat & Care:
Quarter Horses thrive in open pastures with shelter from the elements. They require ample grazing, clean water, and daily handling to maintain their conditioning and temperament. Their hardy constitution suits a range of climates, from the humid South to the arid plains.

Behavior & Temperament:
Gentle, loyal, and highly trainable, the Quarter Horse is known for its calm demeanor and strong bond with handlers. It is alert yet steady under pressure, making it ideal for both work and recreation.

Agricultural Uses & Products:
Renowned for versatility, the breed excels in herding, fieldwork, and riding disciplines, serving as a trusted partner to generations of farmers and ranchers.

Cultural & Practical Notes:
Beloved across the American heartland, the Quarter Horse endures as a living emblem of hard work, partnership, and the quiet dignity of rural life.

APPALOOSA

Equus ferus caballus

Identification:
The Appaloosa is a medium-sized horse distinguished by its striking spotted coat, which may range from subtle speckles to bold, leopard-like patterns. Standing between 14 and 16 hands high, it possesses a well-muscled frame, strong legs, and a broad chest suited for endurance and versatility. Common traits include mottled skin around the muzzle and eyes, striped hooves, and a sparse mane or tail. Each Appaloosa's pattern is unique, a living canvas of nature's artistry.

Origin & History:
The Appaloosa's roots lie with the Nez Perce people of the Pacific Northwest, who carefully bred these horses for speed, stamina, and distinctive color. Their name derives from the Palouse River region where the breed flourished. Following hardships in the 19th century, devoted breeders revived the line, preserving both its heritage and beauty. Today, the Appaloosa stands as one of America's most recognizable native horse breeds.

Habitat & Care:
Adaptable and hardy, Appaloosas thrive in varied climates and open pasture environments. They benefit from regular grooming to highlight their patterned coats and require shelter from extreme weather. Moderate feed, daily movement, and human interaction keep them in peak condition.

Behavior & Temperament:
Intelligent and loyal, the Appaloosa is known for its bold spirit and willingness to work. It forms strong bonds with riders and fellow horses, combining alertness with an easy-going nature.

Agricultural Uses & Products:
Historically prized for cattle work and transport, Appaloosas today excel in riding, showmanship, and leisure use on small farms.

Cultural & Practical Notes:
The Appaloosa remains a vivid emblem of the American West — a testament to the harmony between strength, endurance, and untamed beauty in rural life.

MORGAN HORSE

Equus ferus caballus

Identification:
The Morgan Horse is a compact, finely built breed with an expressive head, arched neck, and strong, elegant carriage. Typically standing between 14 and 15 hands high, it shows balanced proportions and well-defined musculature. Coats vary from deep bay and chestnut to black or palomino, often with a lustrous sheen that enhances its regal appearance. The breed's proud posture and animated stride make it instantly recognizable in motion.

Origin & History:
The Morgan traces back to a single foundation stallion named Figure, later known as "Justin Morgan," foaled in Vermont in the late 18th century. Revered for his strength, speed, and gentle disposition, he became the ancestor of all Morgans. The breed served early American farmers as both workhorse and carriage horse, prized for versatility and stamina. Its influence can be seen in many later American breeds, cementing the Morgan's place as one of the nation's oldest and most iconic horses.

Habitat & Care:
Morgans adapt well to various climates, thriving on moderate pasture and routine care. They benefit from daily exercise, clean stabling, and regular grooming to maintain their sleek coats and muscle tone.

Behavior & Temperament:
Intelligent and affectionate, Morgans are eager to please and form lasting bonds with handlers. They exhibit calmness in harness and vitality under saddle, balancing strength with gentle spirit.

Agricultural Uses & Products:
Traditionally used for plowing, transport, and riding, Morgans remain valued for their versatility — equally at home in farm work, pleasure riding, and showmanship.

Cultural & Practical Notes:
The Morgan stands as a living symbol of American endurance and grace — a reminder of the quiet partnership between humans and horses that built a nation.

SHETLAND PONY

Equus ferus caballus

Identification:
The Shetland Pony is a small but powerfully built horse, rarely exceeding 42 inches in height. Despite its diminutive stature, it possesses remarkable strength and endurance, with a deep chest, short legs, and dense winter coat that protects it from harsh climates. Coat colors include black, bay, chestnut, gray, and roan, often accented by thick manes and tails. Its compact body and intelligent expression reflect centuries of adaptation to rugged island life.

Origin & History:
Native to the Shetland Islands north of Scotland, the breed evolved under severe weather and scarce grazing, developing hardiness unmatched by larger horses. Historically used to carry peat, pull carts, and later to haul coal in British mines, the Shetland earned respect for its stamina and gentle resilience. As mechanization replaced draft animals, the pony found new purpose in children's riding and as a cherished companion.

Habitat & Care:
Shetlands thrive in cool, open pastures and require little more than adequate shelter and moderate feed. Their thick coats demand regular grooming, particularly in warmer climates. Though small, they need ample space to roam and graze.

Behavior & Temperament:
Bright, curious, and occasionally willful, the Shetland is both affectionate and independent. It responds best to patient handling, forming loyal attachments once trust is earned.

Agricultural Uses & Products:
Originally valued for labor, Shetlands today serve primarily as companion animals and in light cart work or children's riding schools.

Cultural & Practical Notes:
The Shetland Pony endures as a symbol of strength in miniature — a testament to nature's ability to refine toughness, beauty, and spirit into one enduring farmyard treasure.

ARABIAN HORSE

Equus ferus caballus

Identification:
The Arabian Horse is a refined and graceful breed, known for its arched neck, dished face, and high-carried tail. Standing between 14 and 15 hands, it combines light bone with remarkable strength and endurance. Coat colors include bay, gray, chestnut, black, white, and roan, often gleaming with a silken sheen. Its expressive eyes and proud carriage give it an unmistakable elegance, equally suited to desert or pasture.

Origin & History:
Among the oldest and most influential horse breeds in the world, the Arabian originated on the Arabian Peninsula thousands of years ago. Bedouin tribes selectively bred these horses for loyalty, speed, and stamina, often keeping them within their tents as cherished family members. The breed's bloodlines spread through trade and conquest, shaping nearly every modern light horse, from Thoroughbreds to Quarter Horses. Revered as the foundation of endurance and beauty, the Arabian remains a symbol of nobility and partnership.

Habitat & Care:
Arabians thrive in open pastures with ample exercise and shelter from extreme cold. Their desert heritage grants them efficiency with feed and resilience in hot, arid climates. Regular grooming accentuates their fine coats and builds trust with handlers.

Behavior & Temperament:
Intelligent, sensitive, and affectionate, Arabians form deep bonds with humans. They are quick learners with a spirited but willing nature, requiring gentle, consistent handling.

Agricultural Uses & Products:
Though once prized for war and long travel, today Arabians excel in endurance riding, showing, and pleasure work.

Cultural & Practical Notes:
The Arabian Horse endures as a living legend — a creature that embodies grace, courage, and the timeless connection between humanity and the horse.

BELGIAN DRAFT HORSE

Equus ferus caballus

Identification:
The Belgian Draft Horse is among the largest and most powerful of all horse breeds, often standing 16 to 18 hands tall and weighing well over a ton. It is broad and muscular with a deep chest, short back, and thick legs adorned with light feathering. The most common coloration is sorrel or chestnut with a flaxen mane and tail, though bays and roans also appear. Despite its massive size, the Belgian carries an air of calm strength and quiet dignity.

Origin & History:
The breed traces its ancestry to the Great Horse of medieval Europe, bred in Belgium's Brabant region for agricultural and war purposes. Over centuries, careful selection produced a reliable draft animal capable of pulling immense loads and working long hours. Imported to America in the late 19th century, the Belgian quickly became a favorite on Midwestern farms for its unmatched power and cooperative nature.

Habitat & Care:
Belgians thrive on ample pasture and shelter from extreme heat. Their size demands abundant feed, clean water, and sturdy stabling. Regular hoof care and conditioning maintain their health and balance.

Behavior & Temperament:
Gentle, intelligent, and eager to please, Belgians are known as "gentle giants." They form strong bonds with handlers and work steadily without complaint.

Agricultural Uses & Products:
Traditionally used for plowing and hauling, Belgians remain valued in forestry, farm work, and ceremonial parades. Their sheer strength and even disposition make them ideal draft companions.

Cultural & Practical Notes:
A symbol of agricultural endurance, the Belgian Draft Horse reminds us of the quiet partnership between human will and animal strength — steady, patient, and timeless.

MUSTANG

Equus ferus caballus

Identification:
The Mustang is a medium-sized, sure-footed horse with a compact, muscular build and a hardy constitution. Standing between 13 and 15 hands, it varies widely in coloration — bays, chestnuts, grays, blacks, paints, and duns all appear within the breed. Its thick mane and tail, strong legs, and deep chest speak to generations of survival in harsh, open landscapes. Though smaller than many domestic breeds, the Mustang's endurance and balance are exceptional.

Origin & History:
Descended from Spanish horses brought to the Americas in the 1500s, the Mustang is a living relic of early exploration and frontier life. As some escaped or were released into the wild, they formed self-sustaining herds across the western plains. Over centuries, natural selection forged a horse of remarkable toughness, blending Iberian grace with feral resilience. Mustangs later became symbols of freedom and the American West, their lineage preserved through both wild populations and conservation programs.

Habitat & Care:
Mustangs adapt well to rugged terrain and sparse grazing, thriving on open pasture. Domesticated individuals require ample space, clean water, and regular exercise. Their hardy nature makes them low-maintenance compared to many purebred horses.

Behavior & Temperament:
Alert, intelligent, and deeply instinctual, Mustangs are quick learners and loyal once gentled. They remain independent-minded, often reflecting their wild ancestry.

Agricultural Uses & Products:
While not traditionally a farm workhorse, Mustangs serve as dependable riding and trail companions, valued for stamina and agility.

Cultural & Practical Notes:
The Mustang endures as a wild emblem of spirit and survival — a reminder that even among the cultivated fields, a trace of untamed freedom still runs in the wind.

PAINT HORSE

Equus ferus caballus

Identification:
The Paint Horse is a striking and athletic breed, distinguished by its bold patches of white and color across a strong, well-balanced frame. Standing between 14 and 16 hands, it combines the muscular build of the American Quarter Horse with the refinement of the Thoroughbred. Its coat patterns — overo, tobiano, or tovero — make every individual unique, with markings that can resemble sweeping brushstrokes across chestnut, bay, black, or sorrel backgrounds.

Origin & History:
The Paint Horse traces its ancestry to the colorful mounts brought to North America by Spanish explorers. Native tribes prized these vividly patterned horses for their beauty, intelligence, and versatility. In the 20th century, breeders formally established the American Paint Horse Association to preserve both the color and the Quarter Horse conformation, producing a breed that embodies the artistry of the frontier and the practicality of the farm.

Habitat & Care:
Paints thrive in open pastures and moderate climates. They require shelter from extreme weather, regular grooming to maintain their glossy coats, and steady exercise to keep their agile bodies in top form.

Behavior & Temperament:
Gentle, intelligent, and highly trainable, Paint Horses are well-suited to both novice and seasoned riders. They are affectionate toward humans and form strong bonds through consistent, calm handling.

Agricultural Uses & Products:
Historically used for ranch work, the Paint's speed and endurance make it ideal for herding, trail riding, and light farm duties. Today, it remains a favorite in rodeos and Western riding events.

Cultural & Practical Notes:
The Paint Horse stands as a living canvas of the American spirit — graceful in motion, loyal in heart, and forever tied to the open land that shaped it.

PERCHERON

Equus ferus caballus

Identification:

The Percheron is a noble and imposing draft horse, renowned for its strength, balance, and grace. Typically standing between 16 and 18 hands high, it has a powerful, broad chest, deep body, and strong legs with light feathering. Most are dapple gray or black, though other colors occur. Despite its massive frame, the Percheron moves with surprising agility and lightness, a reflection of its well-proportioned build and enduring stamina.

Origin & History:

This breed originated in the Perche region of France, where medieval knights once relied on its ancestors as warhorses. Over time, as armor faded from use, breeders refined the Percheron into a versatile draft animal. Imported to America in the 19th century, it became a cornerstone of agriculture and transport, known for its ability to pull heavy loads with calm precision. Today, the Percheron remains one of the world's most admired heavy breeds.

Habitat & Care:

Percherons thrive in cool climates and open pastures. Their large size requires ample feed, access to shelter, and consistent grooming. With proper care, they adapt well to both farm and stable life.

Behavior & Temperament:

Gentle and intelligent, the Percheron is steady under pressure and eager to work. It bonds closely with handlers and shows great patience and confidence in team settings.

Agricultural Uses & Products:

Traditionally used for plowing, hauling, and carriage work, the Percheron remains popular for draft exhibitions, forestry, and light farm labor.

Cultural & Practical Notes:

The Percheron's enduring elegance and power make it a living symbol of cooperation between strength and grace — a quiet reminder of the working heart of rural life.

RHODE ISLAND RED CHICKEN

Gallus gallus domesticus

Identification:
The Rhode Island Red is a sturdy, medium-to-large chicken known for its deep mahogany plumage that glows with copper and crimson tones. It has a rectangular, well-muscled body, a single or rose comb, and bright red wattles. Hens are slightly smaller and lighter in color than roosters, which carry an iridescent sheen across their tail feathers. This breed's hardiness and adaptability make it recognizable on farms worldwide.

Origin & History:
Developed in the late 19th century in Rhode Island and Massachusetts, the breed emerged from crossing Malay, Cochin, and Leghorn fowl. Farmers sought a chicken that could provide both abundant eggs and quality meat, and the Rhode Island Red quickly became an American classic. By the early 1900s, it was recognized as a dual-purpose breed and adopted as the state bird of Rhode Island. Its reputation for reliability made it a mainstay on homesteads and family farms across generations.

Habitat & Care:
These chickens thrive in coops with access to open runs or grassy yards. They tolerate a wide range of climates, though good ventilation and dry bedding are essential. Their dense plumage and vigorous constitution make them ideal for backyard flocks and small farms.

Behavior & Temperament:
Rhode Island Reds are confident, curious, and active foragers. Hens are friendly and steady egg layers, while roosters can be protective but manageable with regular handling.

Agricultural Uses & Products:
Primarily valued for their brown eggs and meat, they remain a cornerstone of sustainable poultry keeping.

Cultural & Practical Notes:
A symbol of dependable productivity, the Rhode Island Red embodies the spirit of the small American farm — humble, resilient, and enduring.

LEGHORN CHICKEN

Gallus gallus domesticus

Identification:
The Leghorn is a sleek, lively breed recognized for its elegant form and glossy plumage. Most are pure white, though buff, brown, and black varieties also exist. They are light-bodied, with long tails, upright combs, and bright yellow legs. Their fine bone structure and sharp alertness give them a refined appearance, while their large, expressive eyes reflect a curious nature.

Origin & History:
Originating from the Mediterranean region of Italy, the Leghorn (named for the port city of Livorno) was imported to America in the early 19th century. Its natural vigor and prolific laying ability soon made it one of the most influential chicken breeds in modern poultry farming. Through selective breeding, farmers refined its egg production while preserving its hardy constitution and adaptability, ensuring its place among the world's foremost laying hens.

Habitat & Care:
Leghorns flourish in well-ventilated coops with ample outdoor access. They prefer mild climates but adapt readily when protected from extreme cold. Their energetic nature means they thrive with space to roam and forage, which keeps them healthy and content.

Behavior & Temperament:
Active and independent, Leghorns are quick, alert, and somewhat flighty. They are not naturally affectionate but remain intelligent and resourceful, always on the move and eager to explore.

Agricultural Uses & Products:
Celebrated for their remarkable egg production, Leghorns lay large white eggs consistently throughout the year. Their lean bodies make them less desirable for meat, but their efficiency has made them a foundation of commercial and backyard flocks alike.

Cultural & Practical Notes:
Graceful and industrious, the Leghorn represents the elegance of simplicity — a tireless layer that has helped feed households and sustain farms for over a century.

PLYMOUTH ROCK CHICKEN

Gallus gallus domesticus

Identification:
The Plymouth Rock is a broad, well-built chicken with a full breast, rounded frame, and calm bearing. Its most recognized variety, the Barred Rock, features alternating bands of black and white feathers that give a charmingly striped appearance. Other varieties include white, buff, and silver penciled. It has a single bright-red comb, yellow legs, and a gentle, confident presence that makes it a favorite among farmers and families alike.

Origin & History:
Developed in Massachusetts in the mid-1800s, the Plymouth Rock was one of the first truly American chicken breeds. Bred from a mix of Dominique and Black Java stock, it quickly became a cornerstone of homestead poultry keeping. Its dual-purpose nature — strong for both meat and eggs — made it invaluable during the early 20th century, especially in wartime America when reliability and thrift were prized on the farm.

Habitat & Care:
Plymouth Rocks are hardy birds that thrive in varied climates. They adapt well to both free-ranging and confined conditions, provided their coops are clean, dry, and well-ventilated. Their thick feathering offers protection against cold weather.

Behavior & Temperament:
Docile, intelligent, and steady, Plymouth Rocks are known for their friendly demeanor. They are excellent layers and attentive mothers, making them a cornerstone of small mixed flocks.

Agricultural Uses & Products:
A dependable dual-purpose breed, they produce a steady supply of brown eggs and tender, flavorful meat.

Cultural & Practical Notes:
The Plymouth Rock remains a symbol of self-reliant farming and American rural heritage — a steadfast bird that reflects the quiet endurance of the family homestead.

SUSSEX CHICKEN

Gallus gallus domesticus

Identification:
The Sussex is a well-rounded, medium-to-large chicken with a broad body, full tail, and gentle expression. Its plumage is most famously white with black neck and tail feathers, though red, speckled, and buff varieties also exist. Each feather is neatly edged, giving the bird a refined and classic look. Its comb is single and red, and its legs are pale pink or white. Sussex chickens are both graceful and practical, embodying a perfect balance of beauty and utility.

Origin & History:
Originating in the county of Sussex, England, this breed dates back to Roman times, making it one of the oldest known chicken breeds. It rose to prominence in the 19th century as one of Britain's leading table fowl, prized for its flavorful meat and reliable laying ability. Over generations, careful breeding preserved its vigor and adaptability, ensuring its enduring popularity on both English estates and small family farms around the world.

Habitat & Care:
Hardy and adaptable, Sussex chickens thrive in most climates. They prefer open yards or grassy runs where they can forage freely but also do well in coops when given adequate space and light.

Behavior & Temperament:
Calm, curious, and sociable, Sussex chickens are known for their gentle nature. They are intelligent foragers and tend to be friendly with people, making them excellent for small farms and backyard flocks.

Agricultural Uses & Products:
A true dual-purpose breed, the Sussex produces rich brown eggs and tender, fine-textured meat.

Cultural & Practical Notes:
The Sussex chicken stands as a living link to traditional English farming — a dignified and dependable bird whose quiet grace continues to enrich rural life.

TURKEY

Meleagris gallopavo domesticus

Identification:
The domestic turkey is a large, broad-breasted bird with impressive plumage and a commanding presence. Males, or toms, display fan-shaped tails, iridescent feathers that shimmer in bronze, green, and copper hues, and distinctive red wattles and snoods. Hens are smaller and more subdued in color, often with brown or gray tones. Modern breeds vary from the traditional Bronze and Bourbon Red to the white-feathered varieties favored in commercial farming.

Origin & History:
Turkeys were first domesticated by Indigenous peoples of Mexico and the American Southwest long before European contact. Early explorers carried them to Europe in the 1500s, where they became a prized delicacy. The bird eventually returned to North America with settlers, now as a domesticated species. Centuries of selective breeding refined the turkey into a reliable source of rich meat, shaping its modern form and size.

Habitat & Care:
Turkeys require spacious pens or pastures where they can roam and forage. They benefit from dry, well-ventilated shelters and access to clean water and grit. Hardy yet sensitive to damp conditions, they thrive best in temperate climates with protection from heavy rain or cold.

Behavior & Temperament:
Curious and intelligent, turkeys are social creatures that communicate with an array of chirps, purrs, and gobbles. Males can be territorial during breeding season but are generally calm when handled with patience.

Agricultural Uses & Products:
Turkeys are raised primarily for their flavorful meat, a staple of festive and family gatherings.

Cultural & Practical Notes:
Long associated with harvest feasts and gratitude, the turkey remains a symbol of abundance and rural tradition — a bird whose legacy is woven into both field and folklore.

MUSCOVY DUCK

Cairina moschata

Identification:
The Muscovy Duck is a large, broad-bodied bird easily recognized by its red facial caruncles — the bumpy skin around the bill and eyes. Its plumage ranges from pure white to glossy black with green and purple iridescence, often with striking patches of both. Males are noticeably larger than females, with heavier builds and more pronounced caruncles. Unlike most ducks, the Muscovy's claws and strong legs make it adept at perching in trees and on fences.

Origin & History:
Native to the forests and wetlands of Central and South America, the Muscovy Duck was domesticated long before European contact. Spanish explorers introduced the species to Europe in the 16th century, where it became prized for its lean, richly flavored meat. Over time, selective breeding led to the modern domestic Muscovy, a quiet, hardy bird now found on small farms worldwide.

Habitat & Care:
Muscovies thrive in semi-wild conditions near ponds or streams, where they can forage for insects, grasses, and aquatic plants. They need shelter from extreme cold and access to clean water for bathing. Their calm disposition and minimal noise make them ideal for rural and suburban homesteads.

Behavior & Temperament:
Gentle and intelligent, Muscovies are quieter than other ducks, communicating in soft hisses and trills rather than quacks. They are excellent mothers and efficient foragers.

Agricultural Uses & Products:
Raised primarily for their flavorful, lean meat, Muscovies also contribute pest control benefits through their voracious appetite for flies and mosquitoes.

Cultural & Practical Notes:
Often seen roosting on barn rails or pond edges, the Muscovy embodies a tranquil, pastoral charm — a reminder of the harmony between domestic life and the natural world.

PEKIN DUCK

Anas platyrhynchos domesticus

Identification:

The Pekin Duck is a large, sturdy bird with a bright white plumage, deep yellow bill, and orange feet. Its body is broad and slightly tilted forward, with a full, rounded chest and upright neck. The breed's clear eyes and snow-colored feathers give it a clean, classic look. While most Pekins are pure white, slight cream or buff tones can appear depending on sunlight and age.

Origin & History:

The Pekin Duck traces its ancestry to China, where it was domesticated thousands of years ago for meat and eggs. In the mid-1800s, breeding stock was brought to Europe and America, where farmers quickly recognized its rapid growth and mild, flavorful meat. The breed became the foundation of the modern commercial duck industry, prized for its consistency and adaptability.

Habitat & Care:

Pekins thrive in grassy enclosures with access to shallow ponds or water troughs for bathing. They are hardy and tolerant of varied climates but should be provided dry bedding and shade in hot weather. Their love of water must be balanced with protection from mud and predators.

Behavior & Temperament:

Gentle, lively, and highly social, Pekins enjoy the company of both ducks and humans. They are alert foragers and communicate with cheerful, nasal quacks that often fill the barnyard.

Agricultural Uses & Products:

Primarily raised for meat, Pekins are valued for their tender flesh and efficient feed conversion. They also lay large white eggs and can serve as charming farmyard companions.

Cultural & Practical Notes:

From the rice paddies of ancient China to modern homesteads, the Pekin Duck endures as a symbol of domestic abundance — calm, content, and timelessly elegant.

TOULOUSE GOOSE

Anser anser domesticus

Identification:
The Toulouse Goose is a large, heavy-bodied bird with soft gray plumage, lighter underparts, and an orange bill and feet. Its posture is slightly stooped, giving it a dignified, almost contemplative appearance. The breed is broad and deep through the chest, with a low-hanging dewlap under the bill in some strains. Show varieties are often massive and slow-moving, while farmyard types are leaner and more active.

Origin & History:
Originating in southern France near the city of Toulouse, this goose descends from the wild Greylag and was prized by farmers for its calm nature and superior fattening qualities. It gained prominence in the 18th and 19th centuries as a key source of foie gras and down. The Toulouse later spread across Europe and to North America, where it became a cornerstone of domestic goose breeding.

Habitat & Care:
Toulouse Geese prefer open pastureland with access to clean water for bathing and grazing. They are hardy in cool, temperate climates but need dry shelter and soft bedding during wet or freezing weather. Ample grass and grain keep them content and productive.

Behavior & Temperament:
Gentle and deliberate, Toulouse Geese are among the most placid of domestic breeds. They are intelligent, slow to startle, and loyal to their mates, making them easy to manage on family farms.

Agricultural Uses & Products:
Raised for both meat and liver production, they are also valued for their rich, insulating down.

Cultural & Practical Notes:
A symbol of pastoral abundance, the Toulouse Goose remains a fixture of European farm life — serene, steady, and steeped in agricultural tradition.

EMBDEN GOOSE

Anser anser domesticus

Identification:
The Embden Goose is a striking, snow-white bird of impressive stature, known for its long, graceful neck and proud carriage. Its bright orange bill and legs stand out against pure white plumage, and its eyes are a clear, icy blue. Strongly built and upright in posture, the Embden combines elegance with power. Males are typically larger and more muscular, while females carry a softer, rounder form.

Origin & History:
The breed originated in northern Germany, near the town of Emden, where it was refined from local white geese over several centuries. Brought to England in the early 1800s and later to America, the Embden quickly became prized for its large frame, fast growth, and mild-flavored meat. It has long served as one of the foundation breeds in commercial goose production.

Habitat & Care:
Embden Geese flourish in open pasture with access to shallow water for bathing and foraging. They are hardy in most climates but prefer dry shelter in winter and shaded resting areas during the heat of summer. With ample grass and grain, they maintain excellent condition and steady growth.

Behavior & Temperament:
Generally calm yet alert, the Embden is less vocal than some other geese but protective of its flock. They are intelligent grazers, forming close pair bonds and displaying quiet affection toward familiar caretakers.

Agricultural Uses & Products:
Primarily raised for meat, Embdens also provide high-quality down and feathers.

Cultural & Practical Notes:
A symbol of purity and pastoral strength, the Embden Goose has long graced barnyards and meadows — a living emblem of quiet dignity and rural prosperity.

GUINEA FOWL

Numida meleagris

Identification:
The Guinea Fowl is a medium-sized, round-bodied bird with a speckled gray-and-white plumage and a bare, bony head crowned with a small horn-like casque. Its wattles are bright red, and its face carries shades of blue and ivory. The bird's strong legs and compact build give it a lively, alert presence. While the "pearl" variety is most familiar, there are also lavender, white, and pied strains.

Origin & History:
Native to sub-Saharan Africa, the Guinea Fowl was first domesticated by ancient peoples who valued its insect-hunting ability and lean, flavorful meat. Brought to Europe through trade routes long before the age of exploration, it became a favorite among farmers and estate keepers for both its utility and ornamental charm. The species' wild nature remains evident even in domesticated flocks.

Habitat & Care:
These hardy birds thrive in open spaces where they can forage freely. They prefer dry ground and require shelter from cold, wet weather but little else. Though capable of flight, they tend to stay near familiar roosts and benefit from secure nighttime housing to protect against predators.

Behavior & Temperament:
Guineas are highly social and vocal, moving in tight groups and communicating through loud, distinctive calls. While wary and sometimes flighty, they are excellent watchdogs, alerting farmers to any disturbance.

Agricultural Uses & Products:
Known for controlling ticks and pests, Guineas provide excellent natural pest management, as well as meat and small, speckled eggs.

Cultural & Practical Notes:
Their wild beauty and vigilant nature make Guinea Fowl a living echo of the untamed savanna — a spirited guardian of the modern farmstead.

DONKEY

Equus asinus

Identification:
The donkey is a sturdy, sure-footed animal, smaller and stockier than a horse, with long ears, short mane, and a distinctive bray. Its coat ranges from gray to brown, black, or white, often marked by a dark dorsal stripe running from mane to tail. Regional breeds vary in size and build — from small Mediterranean donkeys to the larger American Mammoth Jackstock — yet all share an unmistakable air of patient strength.

Origin & History:
Descended from the African wild ass, donkeys were first domesticated in Egypt and northeastern Africa more than 5,000 years ago. They carried goods across deserts and mountains long before the invention of the wheel, serving as one of humanity's earliest beasts of burden. Through centuries of travel and trade, they spread across the Mediterranean and beyond, adapting to diverse climates and farm needs.

Habitat & Care:
Donkeys thrive in dry, open environments but adapt well to temperate pastures. They require simple shelter, fresh water, and coarse forage like hay and grass. Their hooves, suited to hard ground, must be regularly trimmed on soft terrain.

Behavior & Temperament:
Intelligent and gentle, donkeys are cautious by nature rather than stubborn. They form strong social bonds with herd mates and humans alike, communicating through brays and body language.

Agricultural Uses & Products:
Traditionally used for hauling, plowing, and guarding livestock, donkeys remain invaluable on small farms. Their milk is prized for its delicate quality, and their endurance continues to serve rural communities worldwide.

Cultural & Practical Notes:
Symbol of humility and steadfastness, the donkey endures as a loyal companion to humankind — quiet, enduring, and deeply woven into the rhythm of pastoral life.

MULE

Equus asinus × Equus ferus caballus

Identification:
The mule, born of a male donkey and a female horse, combines the strength and stature of its equine mother with the endurance and sure-footedness of its asinine sire. Its coat ranges widely in color — from bay and chestnut to gray or black — and its long ears, short mane, and narrow hooves mark its hybrid heritage. Mules vary in size from light riding types to powerful draft animals, depending on the horse breed used in crossbreeding.

Origin & History:
Mules have been bred since ancient times, with records of their use dating back to Mesopotamia and Egypt. Valued by farmers, soldiers, and traders alike, they became the backbone of transport and agriculture for centuries. Their hybrid vigor — a blend of intelligence, endurance, and resilience — made them indispensable on farms and battlefields alike. Though sterile, each generation was carefully bred anew, ensuring quality and strength through selective pairings.

Habitat & Care:
Mules thrive in varied climates, from rocky mountain trails to humid farmlands. They require the same care as horses — fresh water, good forage, and sturdy shelter — though they tolerate heat and sparse grazing far better. Their hooves need regular trimming, and they benefit from patient handling and companionship.

Behavior & Temperament:
Renowned for intelligence and calm judgment, mules are cautious but not willful. They are steady workers, loyal to handlers who earn their trust, and rarely panic in difficult terrain.

Agricultural Uses & Products:
Used for plowing, hauling, and transport, mules remain prized for their longevity and strength.

Cultural & Practical Notes:
Long a symbol of hard work and perseverance, the mule embodies the enduring partnership between human ingenuity and nature's design — a quiet marvel of utility and grace.

ALPACA

Vicugna pacos

Identification:
The alpaca is a slender, graceful member of the camel family, standing around three feet at the shoulder and weighing between 120 and 200 pounds. Its long neck, short face, and alert eyes lend it a gentle, curious appearance. The fleece, fine and silky to the touch, comes in shades of white, brown, black, gray, and fawn. Two main types exist — the Huacaya, with dense, crimpy fiber, and the Suri, whose fleece forms long, lustrous locks.

Origin & History:
Domesticated more than 6,000 years ago in the Andes Mountains of South America, the alpaca was cherished by the ancient Inca for its luxurious fiber, once reserved for royalty. Descended from the wild vicuña, it has been selectively bred for fiber quality and temperament. In the modern era, alpacas were introduced to farms across the world, from North America to New Zealand, where they remain a cornerstone of sustainable livestock production.

Habitat & Care:
Alpacas thrive in cool, open pastures with dry footing and simple shelter. They graze lightly, making them easy on the land, and require shearing once a year to prevent overheating. A small herd structure suits their strong social instincts.

Behavior & Temperament:
Gentle, intelligent, and watchful, alpacas are herd animals that communicate through soft humming sounds and expressive postures. Their calm nature makes them favored companions on hobby farms.

Agricultural Uses & Products:
Renowned for their fine fleece, alpacas provide one of the softest natural fibers in the world. Their manure enriches soil without burning plants, adding to their agricultural value.

Cultural & Practical Notes:
Still a living link to Andean tradition, the alpaca embodies quiet endurance and grace — a reminder of the harmony possible between utility and beauty in rural life.

LLAMA

Lama glama

Identification:

The llama is a tall, long-necked member of the camel family, standing about four feet at the shoulder and weighing between 250 and 450 pounds. Its wool is thick yet coarse, varying in color from pure white to deep brown or black, often in striking patches. The face is elongated, the ears slightly curved and alert, giving the animal its distinct, inquisitive look. Compared with the alpaca, the llama is larger and bred primarily for strength rather than fiber.

Origin & History:

Domesticated in the high Andes over 6,000 years ago, the llama was the principal beast of burden for pre-Columbian civilizations. The Inca used it to transport goods through mountain passes and revered it as a sacred animal symbolizing service and endurance. Following European exploration, llamas spread slowly beyond South America and have since become valued additions to farms and ranches worldwide for their versatility and calm nature.

Habitat & Care:

Hardy and adaptable, llamas flourish in cool, dry climates but adjust well to temperate pastures. They require shelter from wind and rain, open grazing space, and regular shearing to maintain healthy fleece.

Behavior & Temperament:

Intelligent and observant, llamas are social herd animals with a strong sense of order. They communicate through body language and soft hums and can serve as protective companions for sheep or goats, warding off predators.

Agricultural Uses & Products:

Llamas provide fiber, meat, and manure, but their greatest contribution is labor — capable of carrying heavy loads across rugged terrain.

Cultural & Practical Notes:

Long esteemed as symbols of endurance and cooperation, llamas remind us of the quiet dignity found in work well done and the ancient bond between humankind and its beasts of burden.

RABBIT (New Zealand White)

Oryctolagus cuniculus domesticus

Identification:
The New Zealand White rabbit is a medium-to-large breed, weighing between 9 and 12 pounds, with a compact, muscular frame and pure white coat of dense, velvety fur. Its ruby-red eyes — a hallmark of albinism — contrast vividly against its snowy pelt. The short ears, straight back, and rounded hindquarters give it a balanced, refined look. While uniform in color, minor differences in size and coat texture occur among regional lines.

Origin & History:
Despite its name, the breed was developed in the United States in the early 20th century through selective breeding of Flemish Giants and other white rabbits. It was prized for rapid growth, fine meat quality, and excellent fur — qualities that made it a staple of both small farms and large-scale production. Its dependable temperament and adaptability soon made it one of the world's most widely kept domestic rabbits.

Habitat & Care:
New Zealand Whites thrive in clean, dry hutches or colony-style enclosures with good ventilation and protection from drafts. They tolerate cool climates well but require shade in summer. A balanced diet of hay, fresh greens, and formulated pellets keeps them healthy and productive.

Behavior & Temperament:
Gentle, intelligent, and docile, these rabbits are easily handled and known for their calm disposition. They form social hierarchies when housed together and communicate through soft thumps or nudges.

Agricultural Uses & Products:
Raised primarily for meat and pelts, they are also kept for laboratory research and as pets. Their efficiency and quiet nature make them ideal for small-scale homesteads.

Cultural & Practical Notes:
The New Zealand White stands as a model of farmstead utility — a humble creature offering nourishment, warmth, and quiet companionship across generations.

The Terran Academy

by Luke Thomas

The Illustrated Guide to African Wildlife - A Vintage Field Companion to 50 Majestic Creatures

The Illustrated Guide to Arctic Wildlife - A Vintage Field Companion to 50 Frozen Wonders

The Illustrated Guide to Common Birds - A Vintage Field Companion to 50 Feathered Friends

The Illustrated Guide to Common Farm Animals - A Vintage Field Companion to 50 Barnyard Favorites

The Illustrated Guide to Common Flowers - A Vintage Field Companion to 50 Beloved Blooms

The Illustrated Guide to Northwestern Wildlife - A Vintage Field Companion to 50 Wild Wonders

The Illustrated Guide to Saltwater Fish - A Vintage Field Companion to 50 Ocean Wonders

The Illustrated Guide to Whales and Dolphins - A Vintage Field Companion to 50 Gentle Giants

The Illustrated Guide to Freshwater Fish - A Vintage Field Companion to 50 Shimmering Species

The Illustrated Guide to Prehistoric Dinosaurs - A Vintage Field Companion to 50 Ancient Reptiles

The Illustrated Guide to Common Butterflies - A Vintage Field Companion to 50 Living Jewels

The Illustrated Guide to Common Fruits - A Vintage Field Companion to 50 Natural Delights

The Illustrated Guide to Australian Wildlife - A Vintage Field Companion to 50 Wonders Down Under

The Illustrated Guide to Common Bugs - A Vintage Field Companion to 50 Tiny Wonders

The Illustrated Guide to Common Vegetables - A Vintage Field Companion to 50 Garden Favorites

The Illustrated Guide to Galapagos Wildlife - A Vintage Field Companion to 50 Living Legends

The Illustrated Guide to Poisonous Plants - A Vintage Field Companion to 50 Deadly Beauties

The Illustrated Guide to Wild Cats - A Vintage Field Companion to 50 Fierce Beasts

The Illustrated Guide to European Wildlife - A Vintage Field Companion to 50 Old-World Wonders

The Illustrated Guide to Asian Wildlife - A Vintage Field Companion to 50 Eastern Marvels

The Illustrated Guide to South American Wildlife - A Vintage Field Companion to 50 Vibrant Creatures

The Illustrated Guide to Desert Wildlife - A Vintage Field Companion to 50 Arid Survivors

The Illustrated Guide to Common Spiders - A Vintage Field Companion to 50 Everyday Arachnids

The Illustrated Guide to Tropical Wildlife - A Vintage Field Companion to 50 Untamed Species

The Illustrated Guide to Common Dogs - A Vintage Field Companion to 50 Faithful Friends

The Terran Academy

by Kyle Madison

The Modest Little Snail - *A Tiny Tale About Humility*

The Busy Little Beaver – *A Tiny Tale About Responsibility*

The Focused Little Fire Ant – *A Tiny Tale About Self-Discipline*

The Truthful Little Coral Snake - *A Tiny Tale About Honesty*

The Thankful Little Hummingbird – *A Tiny Tale About Gratitude*

The Brave Little Honey Badger – *A Tiny Tale About Courage*

The Noble Bald Eagle – *A Tiny Tale About Honor*

The Clever Little Red Fox – *A Tiny Tale About Fairness*

The Steady Little Box Turtle – *A Tiny Tale About Perseverance*

The Beluga's Song – *A Tiny Tale About Communication*

The Thoughtful Little Giraffe – *A Tiny Tale About Perspective*

The Gentle Little Donkey – *A Tiny Tale About Patience*

The Clever Little Gecko – *A Tiny Tale About Innovation*

The Kind-Hearted Pelican – *A Tiny Tale About Generosity*

The Wise Little Forest Elephant – *A Tiny Tale About Memory*

The Busy Little Honeybee – *A Tiny Tale About Cooperation*

The Singing Bluebird's Secret – *A Tiny Tale About Joy*

The Curious Little Racoon – *A Tiny Tale About Respect*

The Ladybug's Lucky Garden - *A Tiny Tale About Kindness and Good Fortune*

Made in the USA
Las Vegas, NV
03 November 2026

57424069R00067